Bibliographic information published by the German National Library:

The German National Library lists this publication in the National Bibliography; detailed bibliographic data are available on the Internet at http://dnb.dnb.de .

Imprint:

Print and binding: Books on Demand GmbH, Norderstedt Germany
ISBN: 9783668290297

This book at GRIN:

http://www.grin.com/en/e-book/339118/isolation-identification-and-characterization-of-some-fungal-infectious

Berhanu Kibemo

Isolation, identification, and characterization of some fungal infectious agents of Cassava in South West Ethiopia

GRIN Publishing

Isolation, identification, and characterization of some fungal infectious agents of Cassava in South West Ethiopia

Department of Biology, College of Natural Sciences, Addis Ababa University, Addis Ababa, Ethiopia.

Abstract: Cassava (*Manihot esculenta*) is a dicotyledonous perennial woody shrub with an edible starchy root, belonging to the botanical family Euphorbiaceous. Apart from its use as human food, cassava products also are popular in international trade under different forms. The crop is economically and nutritionally useful in south western parts of Ethiopia. There is scarcity of information so far on the fungal diseases of cassava in Ethiopia. This study was, therefore, initiated to isolate, identify and characterize some fungal infectious agents of cassava in south western parts of Ethiopia. Samples from organs and the plant tissues were first washed in sterile distilled water and surface sterilized in 1% Sodium hypochloride for one minute and then after in 70% alcohol each for one minute. This was followed by rinsing the plant material in sterile distilled water and allowed to dry on sterile tissue paper. The dried tissues were then cultivated on to PDA and incubated at 25^0C. Streptomycin sulphate antibiotic 0.05mg/ml was used to avoid the bacterial contamination. Water agar (WA) media was used for sporulation and to have monoconidal isolates of fungi. Monoconidial isolates of the recovered fungi were stored on PDA slants in the refrigerator at 4^0C for further studies. Eleven isolates were obtained from samples of five locations. However, among these, only six fungal isolates induced pathogenicity on healthy cassava seedlings and leaves. These fungal isolates which grew on glass slide were identified to the genus level by using light microscope as Cephalosporium AAUCF01, Fusarium AAUCF02, Hendersonula AAUCF03, Aspergillus AAUCF04, Penicillum AAUCF05 and Botrytis AAUCF06. The distribution, infection and severity of Fusarium AAUCF02 and Cephalosporium AAUCF01 were more than the other pathogens.

Key words/phrases: - Bio-control, Cassava, Fungal diseases, Pathogenicity, Production

Table of Contents

Introduction

Cassava (*Manihot esculenta*) is a dicotyledonous perennial woody shrub with an edible starchy root, belonging to the botanical family Euphorbiaceous. It belongs to roots and tuber crops that stores edible material in tuber (Howeler, 2003), which belong to class of foods that basically provide energy in the human diet in the form of carbohydrates. Cassava is native to South America, and it is the most widely distributed and cultivated in different parts of the low land tropics. Three continents, Africa, Asia and Latin America produce large amounts of cassava roots. It is the fourth most efficient crop plant, the most widely distributed and cultivated in different parts of the tropics among the tropical root crops (Amsalu Nebiyou, 2003).

Africa is the largest center of production in cassava (Elias and Mikey, 2001). Cassava was introduced to Africa in the latter half of the 16th century from Brazil to the West coast of Africa (Jones, 1959; Amsalu Nebiyou, 2003) by Portuguese navigators and later to East Africa through Madagascar and Zanzibar (Charoentrath *et al.,* 2004) and now grows widely in sub-Saharan Africa (Shittu *et al.,* 2007). Although the crop is often regarded primarily as a famine reserve, there has been increasing realization in recent years of its value as a high-yielding source of carbohydrates. Cassava was first introduced into Ethiopia by the British (Amsalu Nebiyou, 2003).

Over 500 million people in the tropical world particularly in Africa depend on cassava as one of the major staple foods. In Asia and Latin America, productions are largely used as raw materials for industries, as animal feed or for export markets. Apart from its use as human food, cassava products also are popular in international trade under different forms such as dried chips, pellets, flour and starch, thus contributing to the economy of exporting countries (Eke *et al.,* 2007). The advantage it has over other crops particularly, in many of the developing world is its outstanding ecological adaptation, low labor requirement, ease of cultivation and high yields. It is also widely cultivated because it can be successfully grown in poor soils, under conditions of marginal rainfall. It has the ability to grow with appreciable yield where many other crops would hardly survive (O'Brien *et al.*, 1991).

It has been estimated that cassava farmers, typically resource-poor farmers, lose 48 million tons of fresh root, some 30% of total world production, valued at US$1.4 billion every year to pests, diseases, and post-harvest physiological deterioration (PPD) (FAO 2002). Because of its long cropping cycle, 8-24 months, cassava is exposed to an array of pests, diseases and environmental pressures over a prolonged period of time. Therefore, the use of costly inputs, such as pesticides, over the entire crop cycle is prohibitive and uneconomical for the small or large cassava producer (Cock, 1982). Furthermore, the fact that cassava is most often (traditionally) grown on marginal soils reduces the plant of important growth enhancing nutrients and exposes the crop to additional stresses making it more susceptible to pests attack and plant pathogens leading to more severe crop losses (Catalayud *et al.,* 2002).

Cassava yield losses up to 80% due to rot diseases have been reported (Theberge, 1985). Among cassava diseases, cassava root rot (*Nattrassia mangiferae*) and stem rots (*Spherostibe repens)* are the most important in different parts of West Africa (Wydra and Msikita, 1998; Hillocks and Wydra, 2002). These are among the major constraints of cassava in-ground storage. Root rot, caused by root specific fungi, apart from reducing cassava yield can also reduce the quality of cassava root harvest (Hillocks and Wydra, 2002). A number of fungal diseases of cassava have been reported in some countries of Africa such as republic of Congo, Tanzania, Togo, Nigeria, Uganda and other world parts as follows: *Phytophthora drechsleri* (Booth, 1978; Theberge, 1985); *Sclerotium rolfsii* (IITA, 1990); *Rosellinia nectarix* (Lozano and Booth, 1976; Booth, 1978); *Fusarium oxysporum, Botryodiplodia theobromae, Aspergillus niger, Aspergillus flavus, Fusarium solani* and *Macrophomina phaseolina* (Booth, 1978).

Plant pathogens and some insects cause damage and attack on cassava and there by eventually it ends up with a great loss in yield. Fungi are one of the plant pathogens that cause serious diseases of cassava in field conditions (Catalayud *et al.,* 2002). The crop is economically and nutritionally useful in south western parts of Ethiopia. Although research findings have been reported on fungal diseases of cassava in different African countries (Wydra and Msikita, 1998), there is scarcity of information so far on the fungal diseases of cassava in Ethiopia. This study was, therefore, initiated to isolate, identify and characterize some fungal infectious agents of cassava in south western parts of Ethiopia.

Materials and Methods

Sample collection

Diseased cassava samples were collected from different cassava growing areas in order to isolate and evaluate the effect of environmental factors on the pathogens growth. Different parts of the cassava were taken: root, stem and leaf to observe symptoms of the disease. Diseased organs of cassava were sampled from Hawassa/sidama Humbo/Wolayita, W/abaya/Gamogofa, Areka/Wolayita and Jimma zone. The samples were kept in the refrigerator to isolate the fungal pathogens of cassava.

Isolation of fungal pathogens of cassava

Samples from organs and the plant tissues were first washed in sterile distilled water and surface sterilized in 1% Sodium hypochloride for one minute and then after in 70% alcohol each for one minute (Gesier *et al.*, 2005; Summerell *et al.*, 2006). This was followed by rinsing the plant material in sterile distilled water and allowed to dry on sterile tissue paper (Dhingra and Sinclair, 1993; Aneja, 2005; Gesier *et al.*, 2005). The dried tissues were then cultivated on to PDA and incubated at 25^0C (Roux *et al.*, 2004; Gesier *et al.*, 2005). Streptomycin sulphate antibiotic 0.05mg/ml was used to avoid the bacterial contamination. Water agar (WA) media was used for sporulation and to have monoconidal isolates of fungi. Monoconidial isolates of the recovered fungi were stored on PDA slants in the refrigerator at 4^0C for further studies (Gesier *et al.*, 2005; Summerell *et al.*, 2006). The isolates were designated as AAUCF01, AAUCF02...AAUCF011.

Morphological characterization

Morphological studies were carried using slide culture technique. Mycelia fragments of the isolates were inoculated on to PDA that was on glass slide and incubated at 25^0C for 7 days. Afterward, morphological observations were taken based on colony, conidia and conidiophores morphology and other morphological characters as adopted by Barnett and Hunter, (1998).

The effect of culture media on the mycelial growth of the selected fungal isolates

The effect of media was investigated by growing two isolates, AAUCF01 and AAUCF02 on four different media, Potato Dextrose agar (PDA), Potato Sucrose agar (PSA), Malt Extract agar (MEA) and Czapek- Dox agar (CDA). The pH of each medium was adjusted to 3.5. Mycelial fragments were cut off from 7days old culture and inoculated for each medium. Each medium was with three replications and kept in incubator at 25^0C temperature. After 7 days of incubation, the diameter of the mycelia were measured and recorded.

The effect of temperature on the mycelial growth of the selected fungal isolates

The effect of temperature was investigated for the two isolates on MEA by adjusting the incubator in five levels, 15^0C, 20^0C, 25^0C, 30^0C and 35^0C. Mycelial fragments were cut off from 7 days old culture of PDA (Oxoud, pH 5.4 ± 0.2) grown fungal isolates and inoculated on MEA (Oxoud, pH 5.4 ± 0.2). Three replications were prepared for each isolates. After 7days of incubation, the diameter of the fungal mycelial were measured and recorded.

The effect of pH on the mycelial growth of the selected fungal isolates

The effect of pH was studied on potato dextrose broth at 25^0C temperature. The broth pH was adjusted to five levels, 1.5, 2.5, 3.5, 4.5 and 5.5. There were three replications for each pH value. Mycelial fragments were cut off from 7 days old culture and inoculated on potato dextrose broth, and all these treatments were put on to rotary shaker operating 120 rpm and incubated for 10 days. After 10days of incubation, the mycelial biomass was harvested from the broth by using Whatman No. 42 filter paper and dried in an oven under at 65^0C for 48 hours. Thereafter, the mycelia dry weight was measured and recorded.

The effect of light on the sporulation of the selected fungal isolates

The effect of light on the sporulation of the fungal isolates was evaluated while growing the isolates under full dark and light for three different durations with the same light intensity (wave length 300-380 nm, near UV and referred as black light) and 25^0C. Mycelial fragments were cut off from 7days old culture grown on PDA and inoculated on MEA. Thereafter, plates of the fungal cultures were covered with black polyethylene bag to prevent entrance of light (for controls) and the treatments were without cover. Spore counting was done for three times using Heamocytometer.

Pathogenicity tests

The two cassava cultivars, QULLE and KELLO were obtained from Holleta Agricultural Research Center for pathogenicity test with the fungal isolates. Fresh, healthy organs (stem and leaf) from 8-12 month old cassava seedling were washed in running tap water to remove soil and other debris from the surfaces. The organs were surface sterilized in a 1% solution of sodium hypochlorite by immersing them for 1 minute after which they were rinsed with sterile distilled water and left to dry under a laminar flow hood for 3 minutes (Gesier *et al.*, 2005; Summerell *et al.*, 2006). Stems were bored at two edges (upper and lower parts) to a depth of 1cm, using a flame-sterilized 8mm diameter cork borer. A disc of 7 days old PDA culture of each isolate was washed into sterile beakers with sterile distilled water and 1ml of each isolate was inoculated into the hole and sealed with the stems piece removed from the hole. The points of inoculation were sealed with par film to prevent entry of external contaminants as reported by Firdous *et al.*, (2009).

The same procedures were used for the control except that 1ml of sterile distilled water was inoculated into each hole made in the stems. Each inoculated stem was kept in the flasks and incubated at 25^0C for 7 days in order to observe and examine diseases development. The same procedures were used for leaf inoculation except that sterilized needle was used to make pin micks (holes) on leaf surfaces as adopted by Shah *et al.*, (2009).

In the case of *in vivo* test, using the above procedures, stem and leafs of cassava varieties were inoculated. Wounds were made on the surface of the root carefully by using sterilized knife. Then, the root was inoculated by drenching 15ml spore suspensions of fungal isolates in to the seedling in polyethylene bags. It was done with three replications for each isolates. And the inoculated seedlings were kept under shade on open air environment for one month. After 7 days of incubation, it was followed up for the manifestation /development of the diseases symptoms as adopted by Shah *et al.*, (2009).

Evaluation and testing of antagonistic activities of T. harzianum and T. viride against the selected pathogens

Dual cultures method

Antagonism of *T. viride* and *T. harzianum* against the test pathogens was studied by dual culture technique as adopted by Rama *et al.* (2000). The selected tested pathogens were grown on PDA for 7 days. The two antagonists were separately grown on PDA for 7 days. Nine millimeter (9mm) disc of the test pathogen culture was transferred to PDA. Similarly, 9mm disc of each antagonist was transferred and put side by side with the test pathogens. They were incubated for 3 days at 25^0C as the antagonists over grow after 3 days of incubation. The Petri dishes containing the PDA inoculated with the tested pathogen alone served as control. Culture diameter measurements were taken for 3 days consecutively beginning after 24hrs of incubation until the two cultures overlap. The percentage growth inhibition of tested pathogens in the presence of *T. viride* and *T. harzianum* were calculated over control. The growth inhibition was calculated by using the formula as adopted by Rama *et al.* (2000):

$$\frac{\text{Growth in control} - \text{Growth in treatment} \times 100}{\text{Growth in control}}$$

Statistical analysis

The statistical analysis of growth characteristics of the isolates at different media, temperature and pH, and mean comparisons of the isolates based on different parameters were conducted using one way ANOVA procedures of SPSS statistical analysis software (SPSS institute Inc., Cary, NC) version 13. Differences between treatments were determined by using Duncan Multiple Range Test (DMRT) with ($P<0.05$).

Results

Isolation, identification and pathogenicity test of the fungal isolates

Eleven isolates were collected from five locations, Hawassa agricultural Center/Sidama, Humbo/Wolayita, Western abaya/Gamogofa, Areka/Wolayita and Jimma zones. However, among these, only six fungal isolates induced pathogenicity on healthy cassava seedlings and leaves (Table2). These fungal isolates which grew on glass slide were identified to the genus level by using light microscope as *Cephalosporium* AAUCF01, *Fusarium* AAUCF02, *Hendersonula* AAUCF03, *Aspergillus* AAUCF04, *Penicillum* AAUCF05 and *Botrytis* AAUCF06.

As shown in Table2, the two cultivars of cassava were infected and colonized by the fungal isolates. Each isolate had different time of the onset of diseases symptom development. The first disease symptom development by *Fusarium* AAUCF02 was observed after eight days of incubation; small dark circular brown spots appeared on the leaves of QULLE variety which gradually coalesced to form large spots leading to blightening leaves. It was observed that the QULLE variety was more susceptible than KELLO to the test pathogens in terms of the onset of disease symptoms. Re-isolation of the test pathogens from diseased seedlings of cassava revealed that the isolated fungal pathogens were the same as the original identified pathogens of cassava those artificially inoculated on to cassava seedlings. However, some isolates such as *Hendersonula* AAUCF03 on stem and *Penicillum* AAUCF05 on leaf didn't bring apparent diseases symptoms *in vitro* test.

Cultural characteristics of the selected isolates

The fungal isolates showed different growth diameter on different media at 25^0C (Table 4). *Cephalosporium* AAUCF 01 showed maximum mycelial growth (85.3±3.7 mm) on Czapek-Dox agar (CDA) and minimum mycelail growth (68.7±2.9 mm) on Potato Dextrose agar (PDA). The maximum mycelial growth (81.6±2.2 mm) and minimum mycelial growth (78.6±2.3 mm) of *Fusarium* AAUCF 02 was observed on Malt Extract agar (MEA) and on Potato Dextrose agar respectively. The isolates were grown on PDA to investigate the effect of temperature on mycelial growth after 7days of incubation (Table 5). There was high growth for the two isolates within the temperature range 25-30^0C. As the temperature went below and above this range, the two isolates showed decrease in growth diameter. They were more sensitive as the temperature went above 30^0C. The maximum mycelial growth for *Cephalosporium* AAUCF01 (86.3±3.7mm) was at 30^0C and it was (82.3±1.5mm) for *Fusarium* AAUCF02 at 25^0C. There was minimum growth diameter at 35^0C for the isolates, *Cephalosporium* AAUCF01 (21.6±.9 mm) and *Fusarium* AAUCF02 (30.3±.9 mm). *Cephalosporium* AAUCF01 was more sensitive than *Fusarium* AAUCF02 as the temperature went 30^0C. The two isolates showed different growth at different pH levels (Table 6). There was better growth for both isolates at 3.5 pH. There was reduction in mycelial dry weight beyond this value. The isolates showed more reduction in mycelia dry weight as the pH became below the 3.5 than above it. *Fusarium* AAUCF02 showed maximum (2.14±.12 gm) and minimum (0.31±.02) mycelial dry weight at pH of 3.5 and 1.5 respectively. *Cephalosporium* AAUCF01 also showed maximum (2.87±.24 gm) and minimum (0.1±.01 gm) mycelial dry weight at the same pH values as that of *Fusarium* AAUCF02. *Cephalosporium* AAUCF01 was more sensitive than *Fusarium* AAUCF02 as the pH went beyond 3.5 (Table 6). The fungal isolates showed different responses to light on different times of incubation (Table 7). *Cephalosporium* AAUCF01 showed direct relation with the increase of light duration. *Cephalosporium* AAUCF01, after 7days of inoculation, showed little sporulation with regard to plates that were covered with black polyethylene bags (dark). However, there was better sporulation on the plates that were exposed to light. There was excellent sporulation after

7days of incubation for *Fusarium* AAUCF02 in dark where as it was little in light. Culturally, it has no aerial hyphae; colony color, white, smooth in front side of the plate and yellow in back side of the plate.

Evaluation and testing of antagonistic activities of T. harzianum and T. viride against the selected pathogens

The result of *in vitro* evaluation of antagonistic activities of the two biological agents (*T. harzianum* and *T. viride*) showed inhibition of mycelial growth of the test pathogens (*Fusarium* AAUCF02 and *Cephalosporium* AAUCF01). *T. harzianum* had rapid growth over the test pathogens, but it did not show any clear inhibition zone. It showed maximum inhibition (61.5%) and minimum inhibition (20.8%) on *Fusarium* AAUCF02 after 24hrs and 48hrs of incubation respectively at 25^0C. *T. harzianum* inhibition activity was nearly consistent (22-23%) on *Cephalosporium* AAUCF01 within 72hrs (Table 9). *T. viride* showed fast growth forming powdery widespread throughout the Petri plates. As that of *T. harzianum*, it failed to develop inhibition zone, and showed maximum inhibition (52.2%) and minimum inhibition (39.2%) on *Fusarium* AAUCF02 in the 3rd and 2nd day of incubation, at 25^0C respectively (Table 9). There was decrease (48.5-43%) in inhibition as the time goes from 1st to 3rd day on *Cephalosporium* AAUCF01 by *T. viride*. Average percent inhibition of *T. viride* was higher than that of *T. harzianum* for both test fungal isolates of cassava.

Discussion

Eleven isolates were isolated from five different cassava growing areas of South western Ethiopia; of which six isolates were found to incite disease infection on the stem cuttings, leaf and seedlings of QULLE and KELLO cassava cultivars. The isolates were identified by using *Illustrated Genera of Imperfect Fungi* manual based on their morphological and cultural characteristics (Barnett and Hunter, 1998). They were identified as *Cephalosporium* AAUCF01, *Fusarium* AAUCF02, *Hendersonula* AAUCF03, *Aspergillus* AAUCF04, *Penicillum* AAUCF05 and *Botrytis* AAUCF06.

Even though it was not reported as causative agent of cassava diseases, *Cephalosporium gramineum* was known to be the causal agent of Cephalosporium stripe of winter wheat (Bockus and Claassen, 1985) and *Cephalosporium acremonium*, the causal agent of stalk rot and black bundle diseases of maize or Acremonium wilt of sorghum (Bandyopadhyay *et al.,* 1987; Hanlin *et al.,* 1978). Late wilt of maize, caused by the fungus *Cephalosporium maydis* (Samra *et al.,* 1963), is one of the most important fungal diseases in Egypt. This disease also has been reported from India (Payak *et al.,* 1970; Ward and Bateman, 1999). However, in this study, *Cephalosporium* AAUCF 01 was identified to the genus level based on morphological and cultural characteristics. The isolate was observed inciting disease symptoms on cassava seedlings and stem cuttings.

Fusarium AAUCF 02 incited wilt on cassava seedlings, and stem cuttings and leaf under incubator at 25 ^{0}C. In the same study, Hillocks and Waller, (1997) reported that, in some areas, total crop losses have been attributed to rot diseases. *Fusarium* species are a significant component of the set of fungi associated with cassava root rot. Yield losses due to root rot average 0.5 to1 ton/ha but losses >3 ton/ha, an equivalent of 15 to 20% yield, often occur. Numerous and diverse species of *Fusarium* were associated with rotted cassava roots in Nigeria and Cameroon (Bandyopadhyay *et al.*, 2006). Of all diseases caused by *Fusarium* on cassava, the most important one is the vascular wilt disease caused by formae speciales of *Fusarium oxysporum.* The *Fusarium* species that are known to infect a diverse group of plants including crops, ornamentals and trees: *F. graminearum, F. culmorum, avenaceum, F. chlamydosporum,* and *F. verticillioides,* which are the most important *Fusarium* species in central Europe and in large areas in North America and Asia, reduced both crop yield and cereal quality (Gamanya and Sibanda, 2001).

Botrytis cinerea, the pathogen of gray mold disease, causes severe damage on vegetables, ornamentals, fruits, and even some field crops throughout the world. The *Botrytis* disease is the most common disease of greenhouse-grown crops (Choi *et al.*, 2008). Grey mould rot, caused by *Botrytis cinerea* is among the most prevalent storage diseases of carrots grown in temperate regions. It was known that some species of *Botrytis* are plant pathogens with a wide host range that causes heavy yield losses in onion, potato, strawberry, table grapes, and the wine industry (De Curtis *et al.*, 1996). In the other study, Elad *et al.*, (2007) observed that *B. cinerea* can infect numerous host plants; after infection and death of host tissues, the fungus can survive and sporulate as saprophytes on the necrotic tissue, or produce long-term survival structures, such as sclerotia. In this study, *Botrytis* AAUCF06 was observed causing infection on cassava crop creating wilt and spots on leaf and canker on stem.

From research works in the other parts of the world, the majority of the species belonging to the genus *Aspergillus* are reported to be saprophytic in nature. Only a few species including *Aspergillus flavus*, *A. parasiticus* and *A. niger* are considered as weak plant pathogens. These fungi generally infect plant hosts wounded by insects or other agents (Geiser and LoBuglio, 2001). *Aspergillus niger* causes black mould rot that occurs primarily on roots that are wounded and maintained at high temperature. In this study, it was observed that *Aspergillius* can infect cassava crop though it was not identified at species level.

In the case of *in vitro* test, some genera as *Penicillum* AAUCF05 on leaf and *Hendersonula* AAUCF03 on stem cuttings didn't show obvious disease symptoms. This indicates that these genera were organ specific in diseases development on the host. Similarly, Ray and Ravi, (2005) have reported that one of the major constraints of cassava in-ground storage is root rot diseases that is caused by cassava root specific fungi reducing the quantity and quality of cassava root harvest. Although *Penicillium* is a common postharvest fungus, most pathogenic infections occur preharvestly during fruit development. The genus *Penicillium* includes about 150 species but only a minor fraction of these cause infection to important plant and processed foodstuffs (Pitt and Hocking, 1997). Several species of *Penicillium* have been indicated to be pathogenic to a number of different plant hosts. These include *P. expansum, P. italicum, P. digitatum, P. solitum, P. viridicatum, P. rugulosum* and occasionally *P. hirsutum* (Pitt, 1991). In Benin and in parts of Nigeria, root rot (*Nattrassia mangiferae*) is reported to be the most important pathogen (Msikita *et al.*, 2005). *Nattrassia mangiferae* (formerly known as *Hendersonula toruloidea*) is a well-recognized plant pathogen causing branch wilt, canker and dieback disease of a wide range of trees, and storage rot of tubers of plants such as yam (Punithalingam and Western, 1989). In this study also, *Hendersonula* AAUCF03 caused disease symptom on cassava seedlings and leaf but not on stem cuttings.

QULLE variety was more susceptible than KELLO in terms of the onset of disease symptom development and that may be due to various host factors and environmental factors that prevailed to pathogens to incite the disease development. The result of *in vitro* and *in vivo* evaluation and testing of six test pathogens on cassava varieties indicated that the higher infection was caused by *Fusarium* AAUCF02 and *Cephalosporium* AAUCF01. For this study, therefore, only the two pathogens of cassava, *Fusarium* AAUCF02 and *Cephalosporium* AAUCF01 were selected based on their pathogenicity, infection and disease development on the host. *Fusarium* AAUCF02 was isolated from Areka/Wolayita, West Abaya/Gamogofa and Humbo/Wolayita, and *Cephalosporium* AAUCF01 was isolated from Humbo/Wolayita and West Abaya/Gamogofa zones (Table2). Their infection severity was high on cassava seedlings in green house as well as leaf and stem cuttings that were incubated at 25^0 C in laboratory under controlled conditions.

The two genera showed variation in growth on different media as substrate is one of the growth factors. *Cephalosporium* AAUCF01 got maximum growth (85.3±3.7 mm) on Czapex Docks agar (CDA) and minimum growth (68.7±2.9 mm) on Potato Dextrose agar (PSA). *Fusarium* AAUCF02 maximum growth (81.6±2.2 mm) was on Malt Extract agar (MEA) minimum growth (78.6±2.3 mm) was on Potato Dextrose agar. The range of diameter was larger on different culture media for *Cephalosporium* AAUCF01. This showed that it was more sensitive than *Fusarium* AAUCF02 for culture media composition.

As growth factor, temperature affected mycelia growth. Their better mycelial growth was managed in the temperature range 25^0C-30^0C.There was maximum growth for *Cephalosporium* AAUCF01 (86.3±3.7mm) at 30^0C temperature. *Fusarium* AAUCF02 showed maximum mycelial growth (82.3±1.5mm) at 25^0C. This is in line with the conclusions of Nelson *et al.*, (1983) that 25 ^{0}C is suitable for growth of *Fusarium* species. In general, *Cephalosporium* AAUCF01 was more sensitive than *Fusarium* AAUCF02 to wards change in environmental factors and culture media. The genera showed better growth at pH of 3.5- 4.5. Mycelial dry weight showed more reduction when the pH value become below the range than above it. *Fusarium* AAUF02 showed maximum growth at 4.5 pH. Similarly, Negash Hailu, (2007) reported that the optimum pH for some isolates *Fusarium* was 4.5. Light affected sporulation of the genera. *Fusarium* AAUCF02 produced more spores under full dark than light. This indicates that light hindered sporulation for this fungus. *Cephalosporium* AAUCF01 produced more spores as the light duration increases.

In vitro evaluation of *T. harzianum* and *T. viride* indicated that they occupied all the spaces and competed for nutrients thereby hindered the growth of *Fusarium* AAUCF02 and *Cephalosporium* AAUCF01. The absence of inhibition zone indicated that the antagonistic mechanism of these species is through competition for spaces and nutrients rather than forming inhibition zone. Similarly, Negash Hailu, (2007) observed that the overgrowth of *T. harzianum* and *T. viride* against *Fusarium xylarioides*. In this study, relatively *T. viride* exhibited higher percent of inhibition than *T. harzianum* and this is correlated with the report that *T. viride* produced antifungal properties that can able to inhibit the plant pathogens (Lin *et al.*, 1994).

Conclusions

In this study, six fungal genera of cassava diseases in Southwest of Ethiopia were reported for the first time. The distribution, infection and severity of *Fusarium* AAUCF02 and *Cephalosporium* AAUCF01 was more than the other pathogens. In general, *Cephalosporium* AAUCF01 was more sensitive than *Fusarium* AAUCF02 to wards change in environmental factors and cultural media. Each genus had different time for the appearance of diseases symptom. Except *Penicillum* AAUCF05 and *Hendersonula* AAUCF03, all genera were causative agents of systematic diseases. *T. viride* exhibited higher percent of inhibition than *T. harzianum* on both test pathogens. From this, it is possible to conclude that *T. viride* is more effective than *T. harzianum* in controlling cassava diseases caused by *Fusarium* AAUCF02 and *Cephalosporium* AAUCF01.

Acknowledgement

The Authors are thankful to all staff members of Holleta Agricultural Research Center for giving cassava seedlings to conduct pathogenicity test. We would like to thank Biology department, AAU for the provision of transport service while travelling to Holleta to bring cassava seedlings. Finally, We are very much grateful to Ms. Nigate Zewiedu, Mycology Lab. Assistant at AAU for her kindly assistance during experimental works.

References

Aerni, P. (2006). Mobilizing science and technology for development: the case of the Cassava Biotechnology Network (CBN), *AgBioForum,* **9 (1):** 1-14.

Afouda, L. and Wydra, K. (1997). Pathological characterization of root and stem pathogens of cassava and evaluation of antagonists for their biological control: DPG working group. Plant Protection in the Tropics and Sub tropics, Berlin. *Phytomedizin*, **27:** 43-44.

Agrios, G .N. (2005). Plant Pathology. Academic Press. New York. 803Pp.

Amgad, A. S. and John, F. L. (2004). *Cephalosporium maydis* is a distinct species in the *Gaeumannomyces-Harpophora* species complex. *Mycologia*, **96(6):** 1294-1305.

Amsalu Nebiyou (2003). Characterization and Divergence Analysis of Cassava Accession at Jimma. Msc Thesis.

Aneja, K. R. (2005). Experiments in Microbiology Plant Pathology and Biotechnology.4th ed. New Age International Pablishers, New Delhi. Pp. 607.

Ayesu-Offei, E. N. and Antwi-Boasiako, C. (1996). Production of microconidia by Cercospora *henningsii* Allesch., cause of brown leaf spot of cassava. *Annals of Botany*, **78:** 653-657.

Baker, E.F. and Cook, R.J. (1974). Biological control of plant pathogens. W.H. Freeman and Co. Santransisco. Pp. 433.

Bandyopadhyay, R., Mwangi, M., Aigbe, S. O. and Leslie, J. F. (2006). *Fusarium* species from the cassava root rot complex in West Africa. *Phytopathology,* **96:** 673-676.

Bandyopadhyay, R., Mughogho, L.K. and Satyanarayana, M.V. (1987). Systemic infection of sorghum by *Acremonium strictum* and its transmission through seed. *Plant Dis.* **71:** 647–650.

Barnett, H. I. and Hunter, B. B. (1998). *Illustrated Genera of Imperfect Fungi.* 4th edn. St. Paul: APS press. Pp. 40-138.

Belisario, A., Forti, E. and Corazza, L. (1999). First report of *Alternaria alternate* causing leaf spot on English walnut. *Plant Dis.* **83:** 696.

Bilgrhmi, K.S and Verma, R.N. (1982). *Physiology of fungi* 2nd edn. Vikas Publishing House PVT Ltd., New Delhi. Pp.194.

Bockus, W.W. and Claassen, M.M. (1985). Effect of lime and sulfur application to low-pH soil on incidence of *Cephalosporium* stripe of winter wheat. *Plant Dis.* **69:** 576–578.

Bohar, G. and Schwazinger, I. (1999). First report of Septoria species on common ragweed (*Ambrosia artemisiifolia*) in Europe. *Plant Dis.* **83:** 699.

Boher, B., Ptcholo, A., and Tchabama, B. (1997). Identification of the factors enhancing the occurrence of an unusual stem and root rot in cassava in Togo. Cultural practices are accused. *Cahiers Rech.Develop.* **43:**52-58.

Booth, R.H. 1976. Storage of fresh cassava (Manihot esculenta) 1: post-harvest deterioration and its control. *Exper. Agric.*, **12:** 103- 111.

Booth, R.H. (1978). A review of root rots diseases in cassava. In: *Proceedings of 1977 Cassava Protection Workshop.* (Eds.): T. Brekelbaum., A. Bellotti and J.C. Lozano. CIAT, Cali,Colombia. **14:** 121-123.

Burgess, L. W., Summerelle, B. A., Bullock, S., Gott, K. P. and Backhouse, D. (1994). Laboratory manual for *Fusarium* research. University of Sydney, Sydney. Pp.158.

Catalayud, P.A., Polonia, M. A., Seligmann, C.D., and Bellotti, A.C. (2002). Influences of water-stressed cassava on *Phenacoccus herenii* and their associated parasitoids. *Entomologia Experimentalis et Applicata,* **102:** 163-175.

Chadrasekharan-Nair, M., Menon, M.R., Suharban, N. and Verma, A.S. (1979). Anthracnose of cassava: a new record in India. *Curr. Sci.* **48:** 441- 443.

Chang, S. T. (2001). A 40-years journey through bioconversion of lignocellulosic wastes to mushrooms and dietary supplements. Int. *J. Med. Mushroom.* **3:** 209-310.

Charoentrath, S., Boonsang, O. and Narkviot, C. (2004). Biochemical composition in cassava root and physicochemical properties of starch. *Paper presented at the sixth International Scientific Meeting of the Cassava Biotechnology Network, 8-14 March.* Cali: CIAT.

Chet, I. (1987). *Trichoderma* - application, mode of action, and potential as biocontrol agent of soilborne plant pathogenic fungi. In: Chet I (ed). Innovative Approaches to Plant Disease Control. John Wiley, New York, Pp 137-160.

Chevangeon, J. (1956). Le maladie cryptogamique du manioc en Afrique occidental. *Encycopedie mycologique*, **28:** 1-20.

Chijindu, E. N. and Boateng, B. A. (2008). Effect of Nutritional Content of Processed Cassava Chips on Development of Prostephanus truncatus (Horn). *World Journal of Agricultural Sciences,* **4(3):** 404-408.

Choi, G. J., Jin-Cheol, K., Kyoung, S. J., Kwang, Y. C. and Heung, T. K. (2008). Mycoparasitism of Acremonium strictum BCP on Botrytis cinerea, the Gray Mold Pathogen. *J. Microbiol. Biotechnol*, **18(1):** 167–170.

CIAT (1972). Sistemas de producción de yuca. Pp. 99-100. **In:** *Informe Anual.* CIAT, Cali, Colombia.

CIAT (1973). Annual Report 1971. Centro Internacional de Agricultura Tropical, (CIAT), Cali, Colombia.

Cock, J.H. (1982). Cassava: a basic energy source in the tropics. *Science*, **218(45):**755-762.

Cotty, P.J., Bayman, P., Egel, D.S. and Elias, K.E. (1994). Agriculture, aflalatoxin and *Aspergillus* In: The Genus *Aspergillus*. Eds. Powell K.A, Reenwick A & Peberdy J.F. Plenum Press, NewYork. Pp.1-28.

De Curtis, F., Torriani, S., Rossi, E. and De Cicco, V. (1996). Selection and use of Metschnikowia pulcherrima as a biological control agent for postharvest rots of peaches and table grapes. *Ann. Microbiol. Enzimol*, **46:** 45-55.

de Hoog, G.S., Guarro, J., Gene, J. and Figueras, M.J. (2000. Atlas of Clinical Fungi. Centraalbureau voor Schimelcultures, Baarn, and Delft, The Netherlands/Universitat Rovira I Virgil, Reus, Spain. Pp. 442-545.

Denman, S., Kirk, S. A., Brasier, C. M. and Webber, J. F. (2005). In vitro leaf inoculation studies as an indication of tree foliage s usceptibility to Phaytophtora ramorum in U.K. *Plant Pathol.*, **54:** 512-521.

Dhingra, O. D. and Sinclair, J. B. (1993). Basic Plant Pathology Methods. CRC press, Inc. of Baco Raton, Florida. Pp.335.

Drapoch, A.R. (2008). *Cassava as Biofuel Feed Stock.* McGraw – Hill Companies Inc., New York, Pp: 69-78.

Edem, D.O., Ayatse, J.O.I. and Itam, E.H. (2001). Effect of soy protein supplementation on the nutritive value of ' *gari*' (farina) from *Manihot esculenta. Food Chemistry*, **75:** 57–62.

Eke, J., Achinewhu, S. C., Sanni, L., Barimalaa, I. S. and Dixon, M. B. (2007). Seasonal variations in the chemical and functional properties of

Elad, Y., Williamson, B., Tudzynski, P. and Delen, N. (2007). *Botrytis*: Biology, Pathology and control. Published by Springer, Dordrecht, The Netherlands. Pp.412.

Elias, M. and Mikey, D. (2001). Traditional management of Morphological and genetic diversity by the Makushi Amerindians (Guyana, South America) :Perspective for on form conservation of crop genetic resources. *Euphytica,* **120 (1):** 143 – 157.

El-Sharkawy, M.A. (2004). Cassava biology and physiology. *Plant Mol. Biol.* **53:** 621-641.

El-Sharkawy, M.A. (1993). Drought-tolerant cassava for Africa, Asia and Latin-America. *Bioscienc.*, **43:** 441-451.

FAO, (2002). Proceedings of the workshop on processing technology for cassava and other tropical root and tubers in Africa. Abidjan, Ivory Coast, Vol. I and II.

Fasuyi, A. O. (2005). Nutrient Composition and Processing Effects on Cassava Leaf (*Manihot esculenta, Crantz*) Antinutrients. *Pakistan Journal of Nutrition,* **4(1):** 37-42.

Ferdinando, G., Tokeshi, H., Carvhalo, P., Balmer, E., Kimati, H., Cardosa, C. and Salagado, C. L. (1988). Manual de Fitopatologia Doences das plants e seu controle. Biblioteca Agonomica, ceres, sapalo, Brazil.

Firdous, S.S., Asghar, R., Ul-Haque, M.I., Waheed, A., Afzal, S.N. and Mirza, M.Y. (2009). Pathogenesis of Pseudomonas syringae pv. Sesame associated with sesame (*Sesamum indicum*L.) bacterial leaf spot. *Pak. J. Bot.,* **41(2):** 927-934.

Gamanya, R. and Sibanda, L. (2001). Survey of *Fusarium moniliforme (F. verticillioides*) and production of fumonisin B in cereal grains and oilseeds in Zimbabwe. *Int. J. Food Microbiol*, **71:** 145–149.

Geiser, D.M., Lewis Ivey, M.L., Hakiza, G., Juba, J.H. and Miller, S.A. (2005). *Gbberella xylariodes* (anamorph: *Fusarium xylarioides*), a causative agent of coffee wilt disease in Africa, is a previously unrecognized member of the *G. fujikuroi* species complex. *Mycologia.* **97:** 191-201.

Geiser, D.M. and LoBuglio, K.F. (2001). The monophylectic plectomycetes, ascosphaerales, onygenales, eurotiales. **In:** *The Mycota: A Comprehensive Treatise on Fungi Experimental Systems for Basic and Applied Research, Systematics and Evolution.* Eds. D.J Mclaughen, E.G. Mclaughen & P.A. Lenke, Springer-Verlag, Berlin, Germany, VII Part A. Pp. 201-219.

Gilbert, G.S. and DeSteven, D. (1996). A canker disease of seedlings and saplings of Tetragastris panamensis (Burseraceae) in a lowland tropical forest. *Plant Dis.*, **80:** 684-687.

Goh, T.K. and Hanlin, R.T. (1997). Nuclear divisions in the ascus and ascospores of *Mefanospora zamiae. Mycological Research* **101:** 1511-1514. Cloh, T.K. (1999). Single-spore isolation using a hand-made glass needle. *Fungal Diversity,* **2:** 47-63.

Gregory, P. J., Ingram, J.S.I. and Brklacich, M. (2005). Climate Change and Food Security. *Phil. Trans. R. Soc.* **306:** 2139-2148.

Hankoua, B. and Besong, S.A. (2009). Efficient Production of Ethanol from Transgenic CRIS. New York. Pp. 67-71.

Hanlin, T.R., Foudin, L.L., Berisford, Y., Glover, S.U., Jones, J.P. and Huang, L.H. (1978). Plant disease index for maize in the United States, Part I: host index. Agric Expt Sta, Univ of Georgia (Athens), *Res Rept,* **277:** 1–62.

Haran, S., Schinckler, H. and Chet, I. (1996). Molecular mechanism of lytic enzymes involved in the biocontrol activity of *Trichoderma harzianum. Microbiology*, **142:** 2312-2331.

Harman, G.E. (2006). Overview of mechanisms and uses of *Trichoderma* spp. *Phytopathology*, **96:** 190-194.

Harman, G.E. and Bjorkman, P.K. (1998). *Trichoderma* and *Gliocladium* Vol 2. Enzymes, biological control and commercial applications. Taylor and Francis, London, Pp 1-39.

Hillocks, R. J. and Wydra, K. (2010). Bacterial and fungal diseases of cassava. *The American phytopathological society,* **32(8):** 148-150.

Hillocks, R. J. and Wydra, K. (2002). *Bacterial, fungal and nematode diseases.* in:Cassava: Biology, Production and Utilization. R. J. Hillocks, J. M. Thresh, and A. C. Belloti, eds. CABI, UK. Pp. 261-280.

Hillocks, R.J. (2002). Cassava in Africa. **In:** *Cassava: Biology, Production and Utilization*;

Hillocks, R.J., Thresh, J.M., Bellotti, A.C., Eds.; CABI Publishing: Oxon, UK, Pp. 41-54.

Holdenrieder, O. and Kowalski, T. (1999). Pycinidial formation and pathogenicity in *Tubakia dryina. Mycol. Res.*, **92:**166-169.

Howeler, R. (2003). Cassava in Asia: Present Situation and its Future Potential in Agro-Industry. *Journal of Biotech.*, **26:** 45-67.

Hutton, D. G. and Mayers, P. E. (1988). Brown spots of Murcott tangore caused by *Alternaria alternaria* in Queensland. *Australasia Plant Pathol.*, **33:** 49-53.

IITA. (1990). Cassava in Tropical Africa. *A Reference Manual.* Ibadan: Internattional Institute of Tropical Agriculture (IITA). Pp. 176.

Ikotun, T. and Hahn, S. K. (1994). Screening cassava for resistance to the cassava anthracnose isease. *Acta Horticulturae*, **38:** 178-183.

Jones, W. G. (1959). *Manioc in Africa. Palo Alto.* California: Stanford University. Kapoor, I. J. and Kar, B. (1998). Antagonism of Azotobacter and Bacillus against tomato wilt (*Fusarium oxysporum f.sp. lycopersici). Indian phytopathol.*, **42:** 400-404.

Kapat, A., Zimand, G. and Elad, Y. (1998). Biosynthesis of pathogenicity hydrolytic enzymes by *Botrytis cinerea* during infection of bean leaves and *in vitro. Mycological Research.*, **102:** 1017-1024.

Kawano, K. (2003). Thirty Years of Cassava Breeding for Productivity – Biological and Social Factors for Success. *Crop Science,* **43(4):** 1325-1335.

Kay, S. J. and Stewart, A. (1994). Evaluation of fungal antagonists for control of onion white rots in soil box trials. *Plant Path.*, **43:** 371-377.

Kubicek, C.P. and Penttila, M.E. (1998). Regulation of production of plant polysaccharide degrading enzymes by *Trichoderma* in *Trichoderma and Gliocladium: Enzymes, biological control and commercial applications* (Harman, G. E. and Kubicek, C. P., eds) Pp. 49-72, Taylor & Francis Ltd., London, UK.

Laranjeira, D., Santos, E. O., Mariano, R. L. R. and Barros, S. T. (1994). Occurrence of cassava black rot caused by *scyptalidium lignicola* in the state of pernambuco, Brazil. *Fitopatologia Brasileira*, **19:** 466-469.

Leihner, D. (2002). Agronomy and Cropping Systems. *In* Hillocks, R.J.; Thresh, J.M.; Bellotti, A.C. (eds.) *Cassava: Biology, Production and Utilization.* CABI. Pp.47.

Lima, M. F., Takatsu, A. and Reifschneider, F. J. B. (1995). Reaction of cassava genotypes to phytophythora dreschlen. *Fitopatologia Brasileria*, **20:** 406-415.

Lin, A., Lee, T. and Rern, J. C. (1994). Tricholin, a new antifungal agent from *Trichoderma viride* and action in action in biological control of *Rhizoctonia solani*. *J. Antibiot.*, **47(7):** 799-805.

Lozano, J. C. and Nolt, B. (1994). Diseases of cassava (*Manihot esculenta* Crantz). **In:** *Common Names for Plant Diseases*. The American Phytopathological Society, St. Paul, MN. Pp. 36-37.

Lozano, J.C. (1989). Outbreaks of cassava diseases and losses induced. *Trop. Agr. Res. Ser.*, **22:** 113-119.

Lozano, J. C. and Booth, R. H. (1976). Diseases of cassava, 2nd edn. CIAT, Cali, Colombia. Pp. 87.

Makambila, C. and Koumouno, B. L. (1994). The fungal diseases of cassava (*Manihot esculenta* Crantz) in the Republic of Congo, Central Africa. *African Crop Sci. J.*, **2:** 511-517.

McLean, K.L. and Stewart, A. (2000). Application strategies for control of onion white rot by fungal antagonists. *N. Z. J. Crop Hort. Sci.*, **28:** 115-122.

Messiga, A. J. A., Mwangi, M., Bandyopadhyay, R. and Nolte, C. (2004). The status of fungal tuber rots as a constraint to cassava production in the Pouma district of Cameroon. Page 42 in: Book of Abstracts of the 9th Triennial Symposium of the International Society for Root and Tuber crops, Mombasa, Kenya.

Msikita, W., Bissang, B., James, B. D., Baimey, H., Wilkinson, H. T., Ahounou, M. and Fagbemisi, R. (2005). Prevalence and severity of *Nattrassia mangiferae* root and stem rot pathogen of cassava in Benin. *Plant Dis.*, **89:** 12-16.

Msikita, W., Yaninek, J. S., Fiolou, M., Baimey, H. and Fagbemisi, R. (1997). Development of a system to screen and select for resistance to *Fusarium solani*. *Afr. J. Root Tuber Crops,* **2:** 59-61.

Mwangi, M., Bandyopadhyay, R., Dixon, A. G. O. and Tata-Hangy, K. (2004). The status of fungal tuber rots as constraints to cassava production and utilization in eastern

Democratic Republic of Congo. Page 41 in: Book of Abstracts of the 9th Triennial Symposium of the International Society for Root and Tuber Crops, Mombasa, Kenya.

Mwenge, E., Ride, J. P. and Pearce, R. B. (1998). Distribution of Zimbabwean *Armillaria* groups and their pathogenicity on cassava. *Plant Pathology*, **47:** 623-634.

Negash Hailu (2007). Isolation and Identification of *Fusarium xylarioides (gibberella xylarioides)* from Southern Ethiopia, and its Response to Fungal Biocontrol Agents. MSc. thesis. Department of Biology, Addis Ababa University, Ethiopia.

Nelson, P.E., Horst, R.K. and Woltz, S.S. (1983). *Fusarium* diseases of ornamental plants. **In:** *Fusarium: Diseases, biology and taxonomy* (eds. P.E. Nelson, T.A. Toussoun, and R.J. Cook). The Pennsylvania State University Press: University Park and London. Pp. 121-128.

Normanha, S. E. (1970). General aspects of cassava root production in Brazil. **In:** *proceedings 2nd International Symposium on Tropical Root and Tuber Crops*. International Society for Tropical Root Crops/International Institute of Tropical Agriculture, Hawai. Pp. 61-63.

Nweke, F. (1998). The role of root crops in poverty alleviation in Africa. Pp. 102-109. In " Root crops and poverty alleviation''. Proceedings of the 16th symposium of the ISTRC-Africa Branch, Malawi.

Nwokoro, S.O., Orheruata, A.M. and Ordiah, P.I. (2002). Replacement of maize with cassava sievates in starter diets: effect on performance and carcass characteristics. *Tropical Animal Health and Production*, **34(2):** 163-167.

O'Brien, G. M., Taylor, A. J. and Poulter, N. H. (1991). Improved enzymatic assay for cyanogens in fresh and processed cassava. *J. Sci. Food Agric.*, **56:** 277-289.

Oduro, I., Ellis, W.O., Dziedzoave, N.T. and Nimako-Yeboah, K. (2000). Quality of *gari* from selected processing zones in Ghana. *Food Control*, **11:** 297-303.

Okigbo, R.N. and Ikediugwu, F.E.O. (2000). Studies on biological control of post-harvest rot in yams (*Disorea spp*) using *Trichoderma viride*. *J. phytopathol.*, **148:** 351-355.

Onwueme, I. C. (1978). *The tropical tuber crops*. London: John Wiley and Sons Ltd.

Oyeka, T. J. (2005). Cassava root rot fungi in Nigeria: Variability in *Botryodiplodia theobromae* isolates and evaluation of cassava germplasm for root rot resistance. Ph.D. thesis. University of Ibadan, Nigeria.

Papavizas, G.C. (1985). *Trichoderma* and *Gliocladium:* biology, ecology and potential for Biocontrol. *Annu Rev Phytopathol.*, **23:** 23-54.

Patridge, J.E. (1999). Health, disease, and Koch's postulates. **In:** *Inrtrductory plant pathology*. University of Nebraska.Lincoln.http.//plantpath. unl. edu/peartree/homer/sec.skp/1.

Payak, M.M., Lal, S., Lilaramani, J. and Renfro, B.L. (1970). *Cephalosporium maydis*—a new threat to maize in India. *Ind Phytopathol.*, **23:** 562–569.

Peterson, S. W. (2000). Phylogenetic analysis of *Penicillium* species based on ITS and lsu-Rdna nucleotide sequences. **In:** *Integration of Modern Taxonomic Methods for Penicillium and Aspergillius Classification.* Samson, R.A and Pitt, J. I. (Eds). Harwood Academic Publishers, Amsterdam. Pp. 163-178.

Pethybride, S.A., Hay, F.S. and Wilson, C. R. (2004). Short research notes: Pathogenicity of fungi commonly isolated from foliar disease in Tasmanian Pyrethrum crop. *Aurralasian Plant Patho.*, **33:** 441-444.

Pitt, J.I. (2000). Toxigenic fungi: which are important? *Medical Mycology*, **38 (1):** 17-22.

Pitt, J.I. and Hocking, A.D. (1997). *Fungi and Food Spoilage*. 2nd edn. Blackie Academic and Professional, London.

Pitt, J.I. (1991). A Laboratory Guide to Common *Penicillium* species. Commonwealth Scientific and Industrial Research Organization, Food Research Laboratory, N.S.W., Australia.

Punithalingam, E. and Western, J.M. (1989). Descriptions of pathogenic fungi and bacteria No. 274. Hendersonula toruloidea. Kew, UK: Commonwealth Mycological Institute.

Rama, B., Raju, M. and Krishna, M. K. (2000). Efficacy of *Trichoderma* spp. In the management of collar rot of groundnut caused by *Aspergillus niger* (Van Teighem). *Indian J. Plant Prot.*, **28:** 197-199.

Ray, J.D., Burgess, T., Malajcczuk, N. and Hardy, G.E. (2005). Short research notes: First report of *Alternaria* blight of *Paunia* spp. *Australsian Plant pathol.*, **34:** 107-109.

Rebordinos, L., Cantoral, J.M., Prieto, M.V., Hanson, J.R. and Collado, I.G. (1996). The phytotoxic activity of some metabolites of *Botrytis cinerea*. *Phytochemistry*, **42:** 383-387.

Robin, G.W. (1981). Light, algal pigmentation and photosynthesis. In George J., Halasi-kun (ed). Pollution and Water Resource. Colombia University Seminar series. Pollution, Castal Biology and Water Resource Selected Reports, Pergamon Press, New York.

Rondón, G. A. and Aponte, A. (1981). Estudio del superalargamiento de la yuca y búsqueda de cultivares tolerantes a la enfermedad. Fondo Nacional de Investigaciones Agropecuarias(FONAIAP), Maracay, Venezuela.

Roux, J., Van Wyk, M., Hatting, H. and Wingfield, M.J. (2004). *Ceratoctstis* species infecting stem wounds on *Eucalyptus grandis* in South Africa. *Plant patho.*, **53:** 414-421.

Samra, A.S., Sabet, K.A. and Hingorani, M.K. (1962). A new wilt disease of maize in Egypt. *Plant Dis. Reptr.*, **46:** 481–483.

Saleh, A. A., Zeller, K. A., Ismael, A.-S. M., Fahmy, Z. M., El-Assiuty, E. M. and Leslie, J. F. (2003). Amplified fragment length polymorphism diversity in *Cephalosporium maydis* from Egypt. *Phytopathology*, **93:** 853-859.

Samuels, G.J. (1996). *Trichoderma*: a review of biology and systematics of the genus. *Mycol Res.*, **100:** 923-935.

Santos, A. A., Melo, Q. M. S., Teixera, L. M. S. and Fukuda, W. M. G. (1988). Occurrence of cassava rust *Uromyces manihotis* in the state of Ceara. *Fitipatologia Brasileira*, **13:** 73-78.

Shah, S.M.A., Yat-ur-Rahman, H., Abassi, F.M., Akhtar, M.A., Rafi, A. and Khan, A. (2009). Resistance characterization of wild relatives of rice in response to fungal blight. *Pak. J. Bot.,* **46(4):** 917-925.

Shittu, T. A., Sanni, L. O., Awonorin, S. O., Maziya-Dixon, B. and Dixon, A. (2007). Effect of genotype on the flour making properties of some CMD-resistant varieties of cassava. *Food Chemistry,* **101:** 1634–1643.

Simbaya, J. (2007). Results of Cassava-Based Feeding Trials in Poultry, Dairy and Pig Production in Zambia. Lusaka: Livestock Development Trust. Pp. 63.

Sivasithamparam, K. and Ghisalberti, E.L. (1998). Secondary metabolism in *Trichoderma* and *Gliocladium*. In: Kubicek CP, Harman GE (eds) *Trichoderma* and *Gliocladium*, vol 1. Basic biology, taxonomy and genetics. Taylor and Francis, London, Pp: 139–191.

Slope, D. B. (1962). *Cephalosporium* stripe disease of wheat. *Plant Pathology*, **11(4)**: 160.

Smith, O. and Onions, A.H.S. (1994). Preservation and Maintaince of Living Fungi. CAB International, UK. Pp. 104.

Strobe, I. G.A., Yang, X., Sears, L., Kramer, R.S. and Hess, W.M. (1996). Taxol from Pestafotiopsis *microspora,* and endophytic fungus of *Taxus waflachiana. Microbiology,* **42:** 435-440.

Summerell, B.A., Gunn, L.V., Bullock, S., Tesoriero, L. T. and Burgess, L.W. (2006). Vascular wilt of basin in Australia. *Australasian Plant Pathol.*, **35:** 65-67.

Sylvester O. Aigbe & S. U. Remison. (2010). Minor root rot pathogens of cassava (*Manihot esculenta* Crantz) in Nigeria. *Archives of Phytopathology and Plant Protection*, **43** (13): 1335-1341.

Taye Mamo (1994). *Cassava in Southern and South Western Ethiopia.* Cali: CIAT.

Theberge, R. L. (ed.) (1985). Common African Pests and Diseases of Cassava, Yam, Sweet Potato and Cocoyam. International Institute of Tropical Agriculture (IITA), Ibadan, Nigeria. Pp. 107.

Varga, J., Juhász, A., Kevei, F. and Kozakiewicz, Z. (2004). Molecular diversity of agriculturally important *Aspergillus* species. *European Journal of Plant Pathology,* **110:** 627– 640.

Verma, M., Brar, S.K., Tyagi, R.D., Sahai, V., Prévost, D., Valéro, J.R. and Surampalli, R.Y. (2007). Bench-scale fermentation of Trichoderma viride on wastewater sludge: rheology, lytic enzymes and biocontrol activity. *Enzyme Microb. Technol.*, **41:** 764-771.

Vonshak, A., Abeliovich, A., Boussibas, A. and Richmond, A. (1982). Production of Spirulin biomass: effect of environmental factors and population density. *Biomass*, **2:** 175-185. 126. Ward, E. and Bateman, G.L. (1999). Comparison of *Gaeumannomyces*- and

Phialophora-like fungal pathogens from maize and other plants using DNA methods. *New Phytol,* **141:** 323–331.

Wydra, K. and Verdier, V. (2002). Occurrence of cassava diseases in relation to environmental, agronomic and plant characteristics. *Agric Ecosyst. Environ.*, **93:** 211-226.

Wydra, K. and Msikita, W. (1998). Overview of present situation of cassava diseases in West Africa. In: Akoroda MO, Ekanayake I (eds) Proceedings of 6th Triennial Symposium of International Society of Tropical Root Crops - Africa Branch (ISTRC-AB), , Lilongwe, Malawi, Pp. 198-206.

Zeigler, R. S., Alvarez, E. and Lozano, J. C. (1983). Characteristics of cassava resistance to superelongation disease (*Elsinoe brasiliensis*). *Trop. Pest Manage*, **29:** 148-158.

Zhihe, C., Qingping, W., Lihong, X., Xiaoyan, Z. and Jumei, Z. (1998). Advance of biocontrol of *Trichoderma* and *Gliocladium*. *J. Microbiol.*, **25(5):** 284-286.

Appendix

Table 2. Identification and pathogenicity test confirmation of the fungal isolates

No	Fungal designation	Kebele	Woreda	Zone	Region	Pathogenicity test	Appearance of disease symptom
1	*Cephalosporium* AAUCF01 **	Delbo	W/Abaya	Gamogofa	SNNPR	+++	After 11days of incubation
2	*Fusarium* AAUCF02 ***	Dubo	Bolosore	Wolayita	SNNPR	+++	After 8 days of incubation
3	*Hendersonula* AAUCF03	04	Sidama	Hawassa	SNNPR	++_S	After 20 days of incubation
4	*Aspergillus* AAUCF04	04	Hawassa	Sidama	SNNPR	+++	After 15 days of incubation
5	*Penicillum* AAUCF05	Delbo	W/Abaya	Gamogofa	SNNPR	++_L	After 17 days of incubation
6	*Botrytis* AAUCF06	Melko	Sekachekoressa	Jimma	Oromiya	+++	After 13 days of incubation

Key:- "***", found in three sites; "**", found in two sites; ++_S, no diseases symptoms on stem; ++_L, no diseases symptoms on leaf and +++, shows diseases symptoms in all cases on both QULLE and KELLO cultivars.

Table 3. Microscopic observation of the selected isolates

Isolates	*Cephalosporium* AAUCF01	*Fusarium* AAUCF02
Conidia	Produced at tip of hyphae, collected & road in shape	Macroconidia slightly curved or bent at the pointed ends and micro conidia oblong or slightly curved, septated
Conidiophores	Swollen, simple & produced successively	Sporodochia
Mycelium	Septated, branched with long common hyphae & form net like structures	Septated, repeatedly branched, thick, form buds like structures and hyphae are interlocked

Table 4. The effect of culture media on mycelial growth of *Cephalosporium* AAUCF 01 and *Fusarium* AAUCF 02 at 25^{0}C, after 5days of incubation (mm)

Media	*Cephalosporium* AAUCF 01	*Fusarium* AAUCF 02
PDA	$68.7^{c}\pm2.9$*	$78.6^{e}\pm2.3$
MEA	$81.3^{ba}\pm2.3$	$81.6^{d}\pm2.2$
PSA	$77.3^{b}\pm1.5$	$80^{ed}\pm2.6$
CDA	$85.3^{a}\pm3.7$	$79^{e}\pm3.5$

*Values mean ± standard error of three replicate; values followed by the same letter are not significantly different ($P<0.05$).

Table 5. The effect of temperature on mycelial growth of *Cephalosporium* AAUCF 01 and *Fusarium* AAUCF02 on MEA after 5days of incubation (mm)

Temperature	***Cephalosporium* AAUCF 01**	***Fusarium* AAUCF02**
$15^0{}_C$	$48.3^{d}\pm3.2$*	$51^{i}\pm3.8$
$20^0{}_C$	$55.3^{c}\pm3.7$	$63^{h}\pm3.5$
$25^0{}_C$	$84.6^{b}\pm3.2$	$82.3^{f}\pm1.5$
$30^0{}_C$	$86.3^{a}\pm3.7$	$76.3^{g}\pm1.2$
$35^0{}_C$	$21.6^{e}\pm.9$	$30.3^{j}\pm.9$

*Values mean ± standard error of three replicate; values followed by the letter are not significantly different ($P<0.05$).

Table 6. The effect of pH on mycelia dry weight of *Cephalosporium* AAUCF01 and *Fusarium* AAUCF02 after 10days of growth on Potato Sucrose Broth (PSB) (gm) at 25^0C

pH	***Cephalosporium* AAUCF01**	***Fusarium* AAUCF02**
1.5	$0.1^{a}\pm.01$*	$0.31^{d}\pm.02$
2.5	$0.23^{a}\pm.04$	$0.45^{d}\pm.03$
3.5	$2.87^{c}\pm.24$	$2.14^{f}\pm.12$
4.5	$2.27^{bc}\pm.38$	$1.97^{e}\pm.2$
5.5	$1.77^{b}\pm.12$	$1.87^{e}\pm.3$

*Values mean ± standard error of three replicate; values followed by the same letter are not significantly different ($P<0.05$).

Table 7. The effect of light on sporulation of *Cephalosporium* AAUCF01 and *Fusarium* AAUCF02 after 7th day, 10th day and 14th day of incubation at 25^{0}C.

No	Isolates	7th day		10th day		14th day	
		Sporu in D	Sporu in L	Sporu in D	Sporu_in L	Sporu in D	Sporu_in L
1	*Cephalosporium* AAUCF01	1.25×10^5	2×10^5	2.5×10^5	3×10^5	3.25×10^5	3.75×10^5
2	*Fusarium* AAUCF02	4.5×10^5	4×10^5	3.5×10^5	2.75×10^5	2.25×10^5	1.5×10^5

D- dark; L- light

Table 8. Cultural characteristics of *Fusarium* AAUCF02 and *Cephalosporium* AAUCF01 on MEA after 7days of incubation at 25^{0}C

Isolate	Colony colour	Substrate colour	Colony dia meter (mm)	Aerial hyhpae
***Fusarium* AAUCF02**	Mycelium extensive and cottony white , smooth in front side of the plate and yellow in back side of the plate	Light brown	81.6	-
***Cephalosporium* AAUCF01**	Purple(edge), black(center), and white(middle) in front side of the plate and gray being brown at center in back side of the plate	Brown	81.3	+

Key: - -: aerial hyphae absent; +: presence of aerial hyphae

Table 9. Antagonistic effect of *T. harzianum* and *T. viride* against *Fusarium* AAUCF02 and *Cephalosporium* AAUCF01 using dual cultures technique on PDA

Day	Test pathogen	*T. harzanium*	*T. viride*
		% inhibition	% inhibition
1	*Fusarium* AAUCF02	46.5	51.5
	Cephalosporium AAUCF01	22	48.5
2	*Fusarium* AAUCF02	20.8	39.2
	Cephalosporium AAUCF01	23	47.7
3	*Fusarium* AAUCF02	34.4	52.2
	Cephalosporium AAUCF01	23	43